Preface

Genetic engineering admires as a method and process of biotechnology and at the same time both a blessing and a curse? Why cursed? This method uses specifically in the genetic material (genome), and thus in the biochemical control processes of living beings (including the plants).

Genetically modified crop plants are referred to as transgenic crop plants. The first approval of such crop plants took place in 1996. Since then, this technique has grown rapidly. For example, in 19 countries, the number of genetically modified crop plants was increased to 148 million hectares in 29 countries, or 10 percent of global agricultural land.

These crop plants are resistant to plant protection products (for example Glyphosate) or toxic to certain harmful insects due to the genetic modification.

The seed from these plants is more expensive than "normal" seed, but promises higher yields, work facilities, income and health benefits. The environment should also be less affected by the use of genetically modified crop plants. From a scientific point of view, a safety for the environment and health is attested. Environmental associations, suppliers of ecologically produced products and some political parties reject this green technology.

All scientific experiments and commercial cultivation, as any release into the environment, are subject to approval. Why, if everything is so harmless and good?

The transfer of genetic material into another plant is not possible under normal conditions, the proviso is that such "genetically modified organisms" (GMOs) could have negative effects on the environment! So not so safe?

For this reason, the legislature always prescribes an authorization procedure when GM plants are released into the environment.

The legal basis for this in Europe is the "EU Directive on the deliberate release of genetically modified organisms into the environment (2001/18)". All EU Member States have transposed them into national law, in Germany with the Genetic Engineering Act.

The story about Willi the European corn borer should explain without much "thumble-jumble" how the green genetic technology works. At the same time, with this story, I would like to give the readers an idea of the application of green genetic engineering. Let every man form his own judgment!

Nature does not need us - but we need nature!
Peter von Tresckow

This book is dedicated to my wife Beate
and my son Dennis

Before I tell about Willi the European corn borer, I have an introduction to the pest science.

Beneficial or pests, that is the question. The decision is what a benefactor is and what a pest is, finally the man makes the decision . What is a beneficial? The question referred to the invertebrate animals, which include the insects, is answered quite simply. The insects which do not cause economic damage, but prevent it by their presence and way of life, are called beneficial. All other insects that cause economic damage through their presence and way of life are called pests.

Beneficial insects are mostly arachnids (*Arachnida*), since this indeed catch insects for food. But do the spiders catch only pests as food? Certainly not. For spiders can not decide which insect the human being describes as an Beneficial insects or pests. Also the other insects, which use their conspecifics as host, is a division into beneficial and pests probably rather not known.

Bees and bumblebees, as well as other flight insects, are counted among the beneficial insects. Their usefulness is the pollination of fruit trees and other crops. The honey production of the bees is also useful for humans.

The ladybugs are also useful. Since they only eat up to 3000 plant lice or spider mites in their larvae. Most of the ground beetles are predatory and a number of ground beetles have specialized in certain beasts of prey. For example, some species of ground beetles like jumping tails, others like ants and their brood.

Some ground beetles nourish themselves from insects eggs and their larvae, and a few ground beetles feed on seeds and grain, which are then the pests again.

The insects designated as beneficial insects are among others

    Rove or predatory beetles (Staphylinidae)
    Soldier beetles (Cantharidae, Malacodermata)
    Earwigs (Dermaptera)
    Lacewings (Neuroptera)
    Hoverflies (Syrphidae)
    Tachinid flies (Tachinidae)
    Assassin bugs (Heteroptera)
    Ichneumon wasps (Ichneumonoidea)
    Gall midges (Itonididae)
    Hornets (Vespa)
    Bee (Apiformes)

The list of pests is further refined by humans. For example, there are agricultural pests, forest pests, storage pests, wood pests and material pests.

The **agricultural pests** include codling moth, aphids, thrips, Colorado potato beetle, smaller tea tortrix, cherry fruit fly, common cockchafer, European corn borer, plum fruit moth, rhododendron cicada, turnip moth, scale insect, gypsy moth, spider mite, grape, walnut fruit flies, raspberry beetle, whitefly, phylloxera.

Forest pests are: steelblue jewel beetle, bark beetles, oak splendor beetle, oak processionary, green oak leaf, spruce spun sawfly, furniture beetles, pine sawfly, aphids, pine beauty, pine looper moth, spruce sawfly, black arches or nun moth, horse-chestnut leaf miner, gypsy moth, Powder post beetle.

Among the **storage pests** include German cockroach, grain beetle, clothes moth, grain beetles, meal moth.

Wood pests are: Common woodworm, house longhorn, Powder post beetle, termites.

The larvae of the bacon-bugs are called **material-pests.**

A pest may not be missing in the list, the carrot fly. The carrot fly (*Chamaepsila rosae*) is a fly of the family of rust fly (*Psilidae*). Synonyms for *Chamaepsila rosae* are *Musca rosae* (Fabricius, 1794), *Psila rosae* (Fabricius, 1794) and *Chamaepsila henngi* (Tompson & Pont, 1994). They belong to the umbelliferae family and is found in carrots, the most important threat that can lead to total failure of the harvest.

Man in primitive times fed on meat, fish, nuts, fruits and seeds. Carrots, corn, potatoes, apples, pears, oats, barley and much more. These plants, as we know these plants today, were not present in primitive times. It was only man who made what we now call crops from the corresponding crops by breeding. The insects in primitive times already lived on the original forms of today's crop plants. An example would be the Colorado potato beetle, whose original food crop was the sting nightshade (*Solanum rostratum*) was associated as the potato plant of the family Solanaceae. These primal forms of the crops were used to feed insects and to feed the offspring. So to maintain the species. This has not changed in millions of years. The man's use of the primal plants by breeding for himself is a great achievement, for man also wants to preserve his species, and for this he also needs food. Today in the 21st century more than ever.

In early agriculture, the majority of the pests were collected, or combated with simple means such as glue rings. Also, injection molding with chemicals was carried out. To control the phylum, about 1904 sulfur carbon was used. The ground injection with sulfur carbon was indeed an effective, but laborious and expensive method for controlling phylloxera. The liquid, slightly evaporating, toxic sulfur carbon with hand injectors was brought into the main root area of infected vines.

One of the best known pests is the cockchafer (Melolontha melolontha), whose newly hatched cattle require four years of development until they are transformed into a sexually mature animal by metamorphosis. During this time, they are fed up with plant roots. In 1911, about 22 million beetles were collected in an area of about 1800 hectares. Many of these collected cockchafer served as chickens, were fed in the pig's poultry and some roasted into the cooking pot and were processed to soup. For the control of the grubs of the cockchafer (Melolontha melolontha), it was recommended to introduce gelatin capsules filled with sulfuric carbon into the soil. By the massive control of the cockchafer (Melolontha melolontha) with the now forbidden insecticide DDT between the beginning of the 1950er years to about 1972 its population has declined strongly.

An insect which also causes great damage is the Colorado potato beetle. The potato beetles can eat whole fields within a short time. In Europe the potato beetle was first sighted in 1877. In Germany, the first finds were also documented for 1877. In 1887 and 1914 large numbers of infected herds arose in Europe. In 1922, the beetle destroyed 250 square kilometers of potatoes in France. Since the potato beetle does not have any natural bodily faults and each female sheds about 1200 eggs and this happens sometimes in two generations a year, everyone can imagine that of the potato plants, without fighting the beetles, there is not much left. By which, I would argue, imprudent use of many chemical agents, the potato beetle has developed resistances against these agents and is increasingly difficult to combat.

In the forest other pests, the bark beetles (Scolytinae), rage. A not yet exhausted scorch can consist of thousands of copies. 50,000 to 100,000 males and females on a medium-sized spruce are by no means unrealistic, and none of the natural enemies of bark beetles can significantly reduce the population of bark beetles. In addition to the natural spruce forests, man has created optimal bark beetle biotopes with extensive spruce stands. Here climatic extremes, such as long periods of heat or drought, winter with a lot of snow-breaking wood, can multiply the bark beetles explosively.

In 1949, a survey of the German economy by the destruction of the animal pests annulled giant sums.

> 100 million RM (German Reichsmark), of grain by the grain weevils,
> 100 million RM, of fruit by the smaller tea tortrix (worm),
> 20 million RM, of wine by different wine pest,
> 50 million RM, of timber by various wood destroyers,
> 100 million RM, milk, meat and animal skins through the bot fly (Oestridae),

In addition, many other pests come, so that a total of a loss of over 620 million German Reichsmark arose annually. At the same time, the question is asked whether a people can afford it? Then comes the warning:

Anyone who leaves them (the pests) undisturbed in their possessions harms not only themselves, but the whole people, by permitting a diminution of the people's wealth and granting the pests a breeding ground from which they are concerned about the property of the neighbors Spread.

The discovery of the insect-destroying effect of various synthetic substances led to a reversal of the entire pest control.

Thus it is described that if the insects only come into contact with very small traces of these preparations, they are killed. The spraying and atomizing agents are based on dichlorodiphenyltrichloromethane (DDT), hexachlorocyclohexane (Hexa) or polyphosphoric acid compounds. Compared to many old remedies, DDT and Hexa have the advantage of almost complete non-toxicity for humans and pets.

So, as described above, really means such as DDT and Hexa were presented or praised. The chemical club was born.

Advertising for pest control
The approved standard-method against damaging insects

Fly-fight with immediately-effect and duration-effect

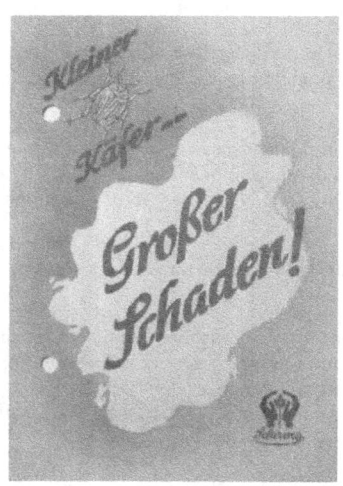

Leaflet for pest control
"Small bug big damage"

"What is of use to me a beautiful garden..."

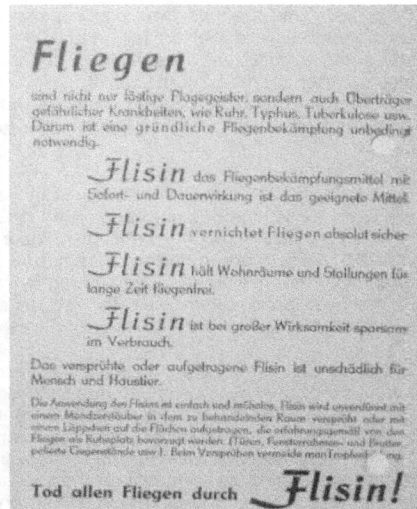

Against aphids and against pests in the soil

Death to all flies!

From slight traces could not be spoken any more. DDT as almost completely nontoxic to humans and to describe the pets is hard to believe. Were there no toxicological investigations at the time?

Toxic   environmental dangerous

GHS Hazardous Substance Labeling for DDT

EU Hazard Identification for DDT

Here are some facts about the use of DDT as a pest control. It was certainly the simple manufacturing process and the good effectiveness against insects that made DDT so successful. The (hypothesized) low toxicity to mammals made DDT decade-long the world's most widely used insecticide. What no one knew at the time (I doubt it) was the fact that DDT is chemically very stable and has a good fat solubility and thus accumulates in the tissues of humans and animals that are at the end of the food chain.

Of course, this has had fatal consequences, as well as has, since still despite a world-wide prohibition of DDT in the year 2004 through the Stockholmer convention. DDT is still being used as an insecticide in countries such as India, North Korea and probably also in other countries in the world, despite DDT's worldwide ban in 2004. Since some degradation products of DDT show a hormone-like effect in the body and DDT is suspected of causing human cancer, the further use of DDT is a crime to humans and nature.

Even before 2004, there was sufficient evidence of the dangers DDT had to face, because the word food chain has not been found in German Duden since 2014. Was and is the person responsible for the profit, the yield on the sale of fruits and vegetables treated with DDT were more important than nature and human health? I cannot imagine that no one knew that insects are used as food by many animals. Birds, frogs, fish and other animals eat insects. People eat fish, storks eat frogs (if still present), but probably the economic factor was the DDT over decades was used.

The production figures of DDT have not been collected and published in all countries. For a long time, the USA was the main producer of DDT, where 74.600 tonnes were produced in 1960. In 1970 it was still 26.900 tons. Only the production data for 1965 are known from the Federal Republic of Germany, at the time it was the second-largest DDT producer in the world with 30.000 tonnes. In the then USSR, between 15.000 and 25.000 tonnes were produced annually in the second half of the 1960s, and in Italy 10.000 tons a year. In the EU countries, about 9.500 tonnes were produced in 1981. For 2005, the world-wide production of DDT was estimated to be 6.269 tonnes of active ingredient divided into India (4.250 tonnes) and China. It is assumed that some 300 tonnes were also produced in North Korea.

Today, other insecticides are used whose long-term effects on the health of man and the conservation of nature are not fully known. Genetic engineering is not the non plus ultra in the control of harmful insects. Here, too, there were incidents. Insects that feed on pollen were eaten by birds, who then simply fell dead from the sky. What happened? The genetically engineered plants had also changed the pollen. While the pollen in the intestinal canal of the insects caused no damage, the birds could not process the pollen absorbed by the insect pollen in the intestine and died.

So, enough of the (long) preface, now I tell the story of Willi.

Willi belongs taxonomically to the family of the Crambidae, also called Crambid Snout Moths. These Crambid Snout Moths are small butterflies. Although they are not as beautifully colored as other butterflies, they are nevertheless quite attractive with a beautiful light-yellow, cream-colored to brick-red coloration in the females and a yellowish-brown, gray-brown to gray coloration in the males. Especially since two transverse lines adorn the wings. The females are slightly larger than the males, but only something. The original habitat is the temperate Europe and Willi will be active when he is grown up in the night. But as far as Willi is not yet.

Willi has not grown up yet. He is not yet a finished insect (Imago), because he is still in the larva stage and for this reason also has a huge hunger for the anger of the farmers. By the way, most people describe this type of larvae as a caterpillar.

The mother of Willi the European corn borer, called Mom

The father of Willi the European corn borer, called Dad

Willi has between 400 and 600 siblings, because his mother has laid many eggs, always in small groups of 10 to 40 eggs. With this large number of brothers and sisters it does not matter to a few more or a few less. There must be so many, since in nature there are enough other living things to feed on insect larvae, and finally, the kind of European corn borer will be preserved.

Willi, of course, does not know his brothers and sisters, perhaps only those who have been taken with him in the same pile and hatched with him. He is concerned primarily with his own well-being.

The life age as an adult European corn borer is by the way not very long, it is only 18 to 24 days. In these 18 to 24 days, it will also be important for Willi to look for a female to look after the next generation.

Well, 18 to 24 days of life is not much, but his larvae time, so the childhood of Willi with more than 50 days is quite long. The larvae time varies considerably and depends on the weather. Especially from the temperature. In short, the colder it is, the longer the development time.

In the course of the late summer, the eggs are placed on the underside of a maize-plants where the eggs are protected from rain and too much sun. After 7 to 14 days Willi and his siblings hatch from the eggs.

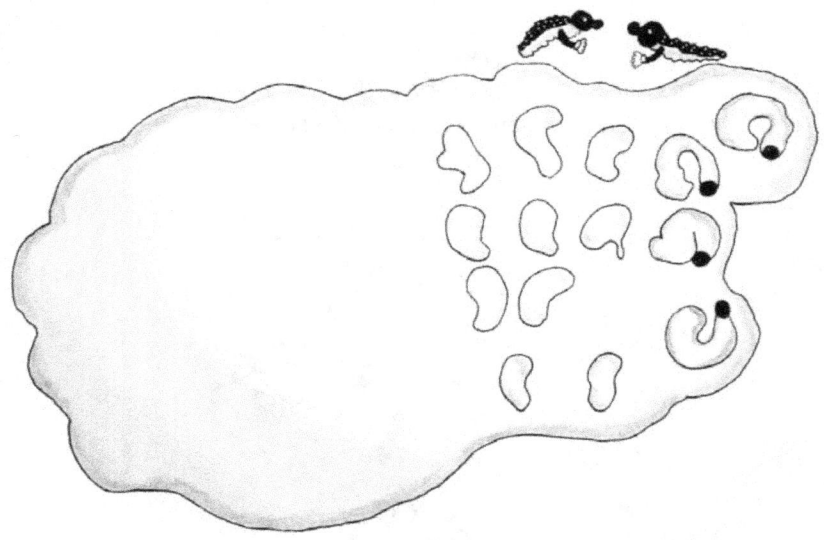

Willi hatched (*right in the picture above*)

The egg hatched from the Willi was hatched by his mom on the leaf of a maize-plant. Other breeds have spread his mama well in botany. About 20 plant species were deposited. Much is on the menu. Genuine hops, potatoes, tomatoes, peppers, fennel, millet, hemp, turnips, buckwheat, real celery and mugwort. Almost nothing is left out. The European corn borer as a vegetable omnivore and gourmet!

This is **Willi** larval ...

... and this is the maize field where Willi lives.

When Willi goes to the inside of the maize stalk, from the bottom of the maize plant, he gets scared. Afraid that his mom hatched the egg he hatched has deposited on a so-called Bt-maize. Willi were given two warnings on his way of life. Once the warning of the Bt maize and once the warning of the parasitic wasp named *Trichogramma brassicae*.

Because this parasitic wasp prefer their eggs in the larvae of the European corn borer and thus provides for their own offspring, which would mean for Willi the end. Thus, Willi quickly sets off on the way to the maize plant and the first danger is averted.

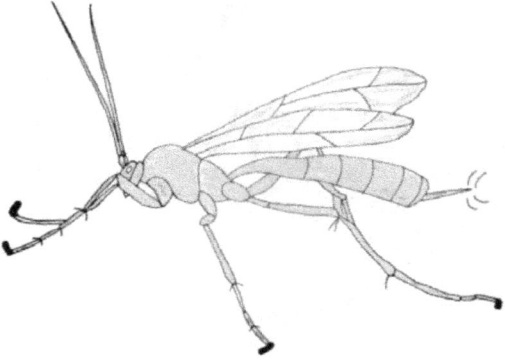

SLIPI the parasitic wasp

But what about his fear of Bt maize. If it is the stem of a Bt-corn plant in which it is, then the eating of this plant is fatal to Willi. What will Willi do? He decides to eat a very small amount of the marrow's stems and waits a long time if he gets sick.

Nothing happened and Willi was lucky. It is not Bt maize or Willi comes from a European corn borer which has developed resistance to the Bt maize.

We let Willi eat in silence and turn to the subject of transgenic maize.

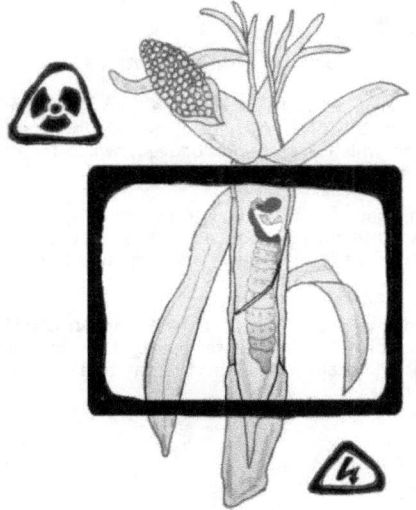

Caution! X-Rays

Genetically modified maize is referred to as transgenic maize (GM maize). In transgenic maize species, certain genes from other organisms are introduced into the maize genome (genotype of a living organism), with the aim of improving the control of harmful insects in Bt maize.

Using the example of a genetically modified maize species, with which one tries to destroy the European corn borer, I will once describe how this genetic engineering is carried out.

The European corn borer is one of the most economically important pests on maize. The FAO estimates that about 4 percent of the annual maize harvest is destroyed by the caterpillars of the European corn borer worldwide by pollinating the flowers and fruit stands as well as the marrow of the stems.

FAO = *Food and Agriculture Organization of the United Nations*

The GM maize is named Monsanto MON 810. To protect the corn before the European corn borer, the soil bacterium is from the DNS (carriers of genetic information) *Bacillus thuringiensis* gene sequence excised which a protein produced, which for the European corn borer is poisonous. The larvae that eat the maize now perish.

Bacillus thuringiensis is a bacterium that is found mainly in the ground, but also in plants and insects in carcasses. The so-called Bt toxins produced by the bacterium are produced during the sporulation (spore formation) of the bacteria and are stored as crystals in the bacteria.

In the gut of a host organism, in our case the larvae of the European corn borer, these crystals are dissolved. This produces proteins which interfere with the metabolism of the insect larvae and virtually destroy it. This causes the insect larvae to die.

Bacillus thuringiensis was first described in 1901 in Japan as Bacillus sotto and 1911 by Ernst Berliner, who gave him the name of *Bacillus thuringiensis* were. In Japan the bacillus was found in silkworms. Ernst Berliner discovered the Bacillus in flour moth. The different subspecies of this bacterium produce over 200 different so-called Bt toxins that are specifically fatal in certain insects.

But as that comes for the European corn borer damaging gene sequence of the bacterium into the cells of the corn-plant? The gene sequence of the harmful properties of the *Bacillus thuringiensis* isolated ("cut"). The gene sequence taken is incorporated into another bacterium. In so-called fermenters (bioreactors) these bacteria are multiplied with the altered gene sequence under optimal nutrient conditions.

After this process, metal microparticles are coated with the modified genetic material and pressed by compressed air onto the plant tissue of the maize plant. Through this process, the cell walls are perforated and the metal microparticles thus reach the interior of the plant cell. The modified genetic material is incorporated into the plant. The maize plant thus manipulated now has the desired property, the so-called resistance to the European corn borer. All Corn-grains naturally also contain this genetic material, so that after sowing every maize plant contains the altered genetic material.

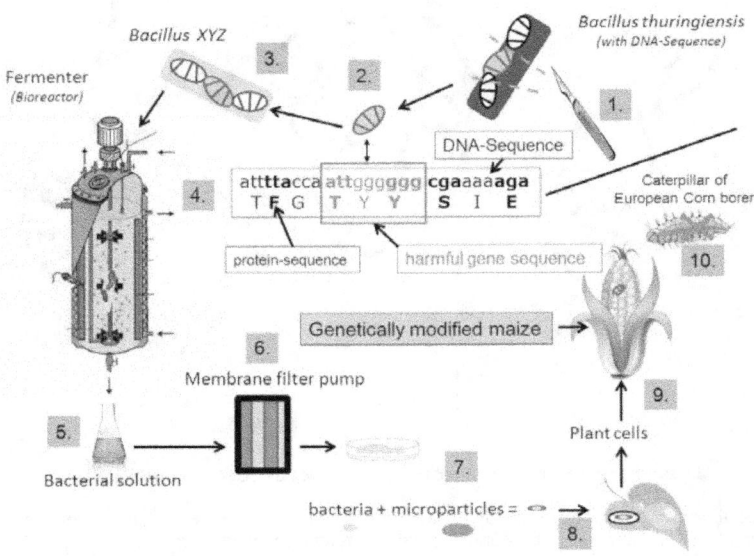

Change of the genome with foreign genes

Of course, it is also said that this genomis is not dangerous for humans and animals. In the meantime, however, it was found that the European corn borer after about 4 years developed resistance to this protein and the genetically modified maize no longer kills the larvae. The prescription of corporations like Monsanto and Co dissolve it. The transgenic Bt maize hardly helps against pests. One of the reasons for this is the unbridled cultivation of monocultures.

The European corn borer is one of the most frequently studied plant pests in the world. Originally living in Europe, from southern Norway to the British Isles, the European corn borer was cosmopolitan (worldwide) through human abduction, and is trapped in North Africa, Asia Minor, West Asia to Turkestan, and North America, where between 1910 and 1920, to find.

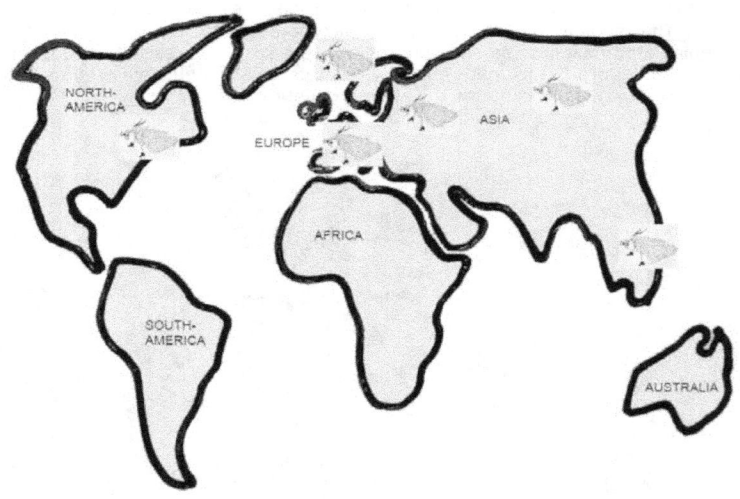

Dissemination of the European corn borer (*Ostrinia nubilalis*)

The production of adequate food for the world's population is very important that is not a question, but not at any cost. Monocultures, eradication of insects, genetic engineering without long-term testing (the emphasis is on long-term) and the impact of genetic engineering in the natural environment can have long-term fatal consequences for humanity.

Every living creature, and if it consists only of a cell, will always try to preserve its kind. Whether it is the potato beetle, the HIV virus, the malarial virus or the bird flu virus.
Each of these creatures will, by adaptation, always try to defend themselves against attacks on their species, for example by resistances or the like. This is faster for the underdeveloped creatures than for higher developed creatures. The more complex creatures have more problems to cope with the adaptation in a short time. The time, as we know, is relative.

Now back to Willi. Willi and his siblings are busy eating. They migrate from maize plant to maize plant and leave broken corn stalks back. Because of the missing mark in the stem the plant lacks the stability. This naturally also makes subsequent harvesting more difficult. The nutrients from the plant root get into the flask through the stem mark. If the marrow is no longer present, this process cannot take place and the corn syrup hardly produces grains. In addition, Willi and his brothers and sisters had a higher degree of disease of the maize plant caused by infection with mold fungi. The affected maize is only suitable for biogas production. It is no longer suitable for grain yield (grain maize) or energy yield (feed maize).

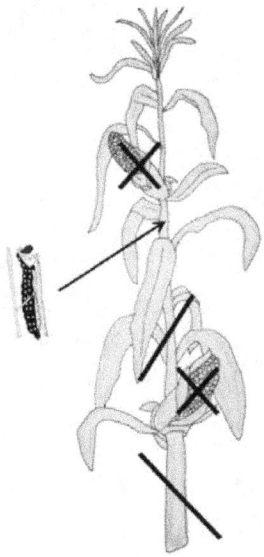

There is also a aesthetic problem. The grub-damage on the corncobs do not lure a buyer for fruit and vegetables. Apples with boreholes from the worm of the apple winder also buys no customer.

Very many fields are planted with maize plants, too many fields. The whole maize plant is chopped at the same time and spent on empty surfaces.

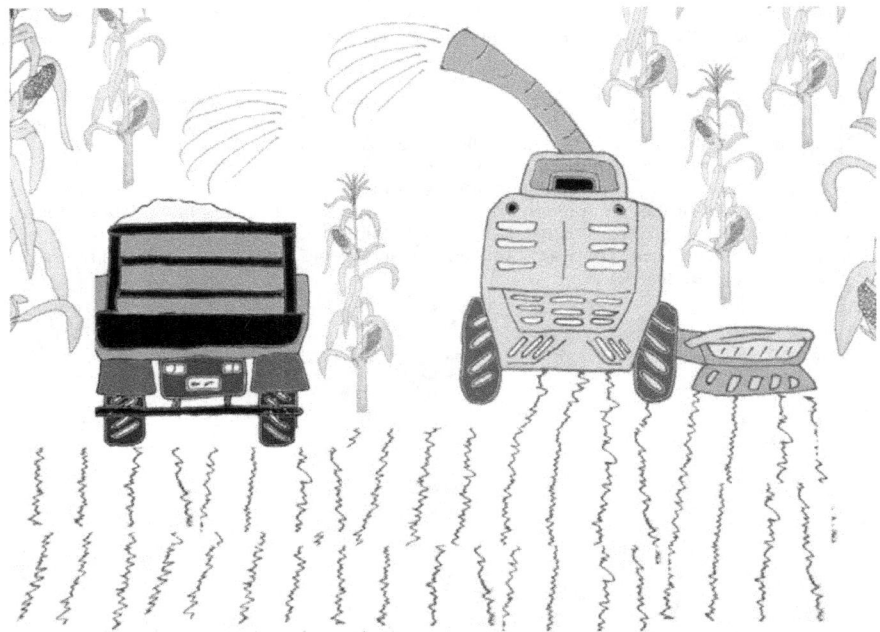

The resulting hills are compacted by tractors. A black film is then drawn over it. The whole serves as a material for the biogas plant.

No grain maize for that silo and no feed-corn for the cows.

When I think about how many people in the world are hungry and the corncobs could have made maize flour for their dumplings to satisfy their hunger, I get angry.

The year is ending and Willi is preparing for the winter. If the maize plants are cut down, the stubble remains on the field. Willi is sitting in one of these stubble and will spend the winter there too. He had luck. Many of his brothers and sisters had no luck, for they were sitting in the mown corn plants and were killed for the most part by the chipping. The larvae, which have survived this procedure die at the latest when compressing the chopper.

This is certainly sad for the single larva, because the larva probably would have liked to live on and founded a family with very many children. For the species itself the loss of the many larvae plays no great role. For nature has, whatever it may be, the species which have many enemies, whether natural enemies, or man with his chemical lobe, a very large number of offspring, so that the species may be preserved. Just to remind **a European corn borer females** lays **400 to 600 eggs** !

I find this fascinating, even if the European corn borer are pests. Where, with the pest claims man. Many birds see it quite differently with the pest. They are happy about many larvae, because these larvae are urgently needed for the rearing of the bird brood. Only a very small part of the birds can feed their brood with grain feed.

In the meantime, it has been spring and Willi has survived the winter in his maize stubble well, because fortunately the winters here in Germany are not as hard as before. It forms a loose cocoon from fine threads and pupated.

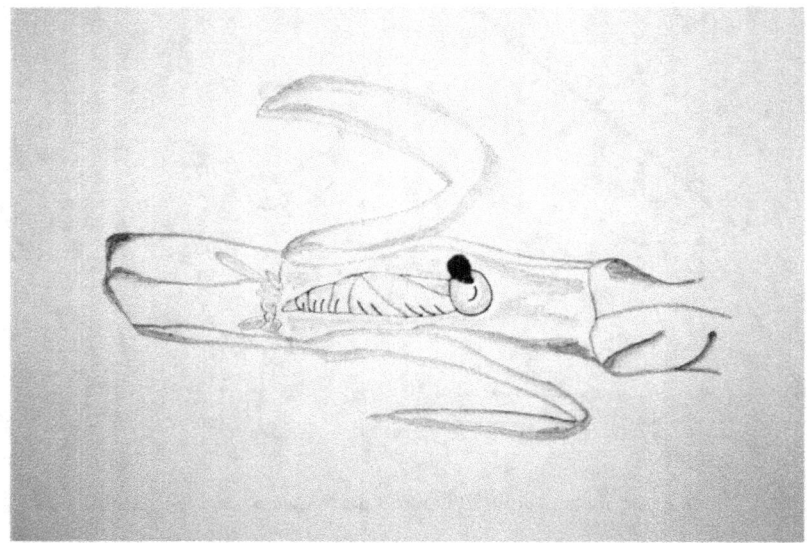

In this chrysalis Willi remains about 12 days. Then he slips and is now grown up. Willi is a finished European corn borer. Now he has 18 to 24 days to look for a female to start his family. Good luck Willi and thank you, I was able to tell your life story.

Alteration of genetic information with foreign genes (larger image)

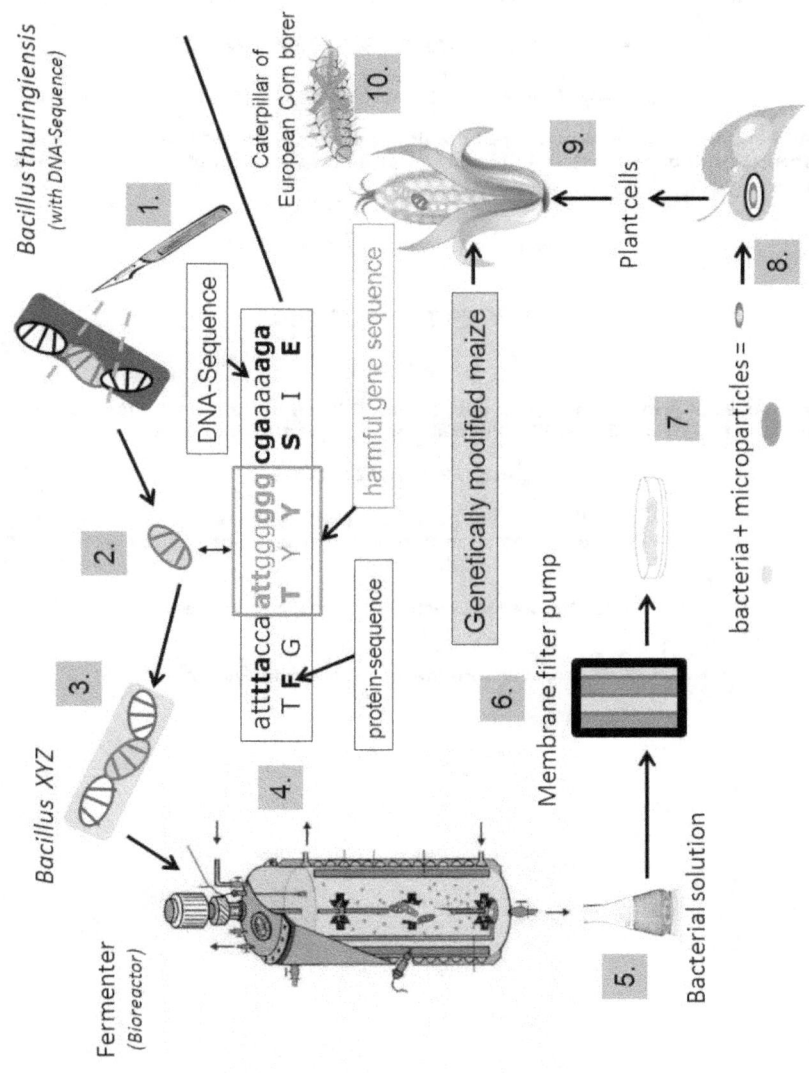

The following books have been published by me and are available in bookshops

FASZINATION INSEKTENWELT

Beetle and Co viewed through the lens

As hardcover bound and printed on high quality paper. Edition in German A5 format, 460 pages, with 849 mostly color illustrations, photos and drawings.

ISBN: 978-3-7357-4283-4

EBook ISBN: 978-3-7357-6785-1

INSEKTENKUNDE

entomology

As Paperback printed on high quality paper. Edition in German A5 format, 400 pages, with 772 black / white illustrations, photos and drawings.

ISBN: 978-3-7412-8985-9

**All books are at the publishing house
BoD - Books on Demand, Norderstedt
Appeared**

These books are written in German language

See also: **www.meine-buchvorstellung.com**

All drawings are by myself (Detlef Schmidt)

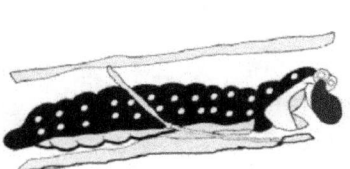

Bibliographic information of the German National Library:
The German National Library records this publication
in the German national bibliography; detailed bibliographic
Data are accessible via http://dnb.dnb.de available.

© Copyright by Detlef Schmidt

Production and publishing:
BoD - Books on Demand, Norderstedt
ISBN: 9783743137066

www.ingramcontent.com/pod-product-compliance
Lightning Source LLC
Chambersburg PA
CBHW071222240526
45470CB00018B/2292